BENT & BATTERED ROTORS

Volume 3

By Wayne Mutza

Color by Don Greer & Tom Tullis

squadron/signal publications

Marines grab their weapons and run from this downed UH-1E Huey gunship that was shot down and crashed into a rice paddy in South Vietnam during the early 1960s

ISBN 0-89747-306-X First Edition

If you have any photographs of aircraft, armor, soldiers or ships of any nation, particularly wartime snapshots, why not share them with us and help make Squadron/Signal's books all the more interesting and complete in the future. Any photograph sent to us will be copied and the original returned. The donor will be fully credited for any photos used. Please send them to:

Squadron/Signal Publications, Inc.
1115 Crowley Drive.
Carrollton, TX 75011-5010

Acknowledgments

Robert L. Lawson-Tailhook
John H. Cline - NASA
Johnny V. Marshall
CAPT K. G. Craig - USN
Carl Vogel
LT Pichard Kyle - USAF
Raymond Fletcher
Mike Sloniker
Larry Davis
Terry Love
Tom Garcia
Dave Fesmire
Dave Grieger
U.S. Navy
Stephen Hayduk
Howard K. Butler
Irwin Baeder
Norman LaFountaine
Jack Calve
CAPT R.E. Joslin - USMC
Bob Steinbrunn
Dennis Dennison
John Konek
Bob Chenoweth
Tom Hansen
Fred Thompson
U.S. Army - Regina Burns
USMC

Dedication

Respectfully dedicated to U.S. helicopter crew members; may they never forget those who served before them, many of whom paid the ultimate price. And to Debbie - my inspiration.

Overleaf: This U.S. Air Force Sikorsky H-5 Dragonfly of the 3rd Air Rescue Squadron was downed over Korea during September of 1951. (Smithsonian Institution)

Introduction

"The thing is, helicopters are different from fixed wing aircraft. A fixed wing aircraft by its nature wants to fly, and if not interfered with too strongly by unusual events or by a deliberately incompetent pilot, it will fly. A helicopter does not want to fly. It is maintained in the air by a variety of forces and controls working in opposition to each other, and if there is any disturbance in this delicate balance the helicopter stops flying, immediately and disastrously. There is no such thing as a gliding helicopter.

This is why being a helicopter pilot is so different from being a fixed wing pilot and why, in general, fixed wing pilots are open, clear-eyed, buoyant extroverts and helicopter pilots are brooders, introspective anticipators of trouble. They know if something bad has not happened, it is about to."

This sobering commentary was cited by Harry Reasoner in February of 1971 and as if to lend credence to his words, the following month was the worst of the Vietnam war for Army helicopters: 143 were lost, 110 to hostile action.

During the early 1940s the highly vulnerable helicopter struggled through its developmental infancy. Frail and under powered, it was limited by range and inability to fly in bad weather. As performance and reliability improved, military planners soon realized the helicopter's potential. The newly invented helicopter was introduced into military service in April of 1943 when four prototype Sikorsky YR-4s were pressed into service direct from the factory to help form the lst Air Commandos. One year later one of these flew deep into Japanese-held territory to rescue a downed pilot and three wounded soldiers.

The onset of the Korean war in June of 1950 fostered the first large scale use of helicopters. When the war began, vintage Second World War helicopters were immediately pressed into service for combat rescue. The first loss of a helicopter on a rescue attempt was an H03S-1 of HU-1, Det. 7 near Sarowan, North Korea on 23 August 1950. The Sikorsky was shot down and ditched in the Han River. In early 1951 the first helicopter transport unit, HMR-161, was formed by the Marines and sent to Korea later that year. By April of 1952 no more than seventy helicopters operated along the front, tasked primarily with medical evacuation and downed aircrew recovery. Nearly 20,000 casualties were evacuated and 996 aircrew were rescued with only slight helicopter losses during the war. Helicopter crewmen became a breed of their own as they ushered in the "airmobile" concept which would reach its fruition in Southeast Asia.

Early helicopters were rather unstable and prone to crashes. This Sikorsky H-5 Dragonfly was destroyed in a crash at Wright-Patterson Air Force Base, Dayton, Ohio during 1947. (via Larry Davis)

A Sikorsky H-5 Dragonfly lies derelict following a crash at Wright-Patterson AFB during 1947. The severed tailboom and three-bladed tail rotor were beneath the fuselage. (via Larry Davis)

In its quest to develop airmobility, the Army had amassed an inventory of 2,753 helicopters by June of 1961. On 2 January 1962 two companies of Army CH-21 Shawnees conducted the first helicopter assault of the Vietnam War. With this assault, the "helicopter war" began in earnest. At the height of U.S. involvement in Southeast Asia, the Army boasted the third largest air force in the world with a peak inventory of 10,017 helicopters, during mid-1970. Nearly 4,000 were in Vietnam and almost 3,000 of those were Bell UH-1 Hueys and its AH-IG Huey Cobra derivative. Throughout the conflict over 10,000 helicopters were committed to the war which brought the helicopter to the forefront as the most valued tool in combat history.

Throughout its growth it distinguished itself as a multi-purpose aircraft with its use limited only by the users imagination. The helicopter had truly come of age but it remained unforgiving when mishaps occurred. Such events often stemmed from a lack of appreciation for what a helicopter could - and could not - do.

Since the employment of helicopters is governed by their aerodynamics, an understanding of the aircraft's basic characteristics was important. At increased altitude and temperature, air becomes thinner severely reducing lift capability. This "density altitude" (prevalent in Southeast Asia) coupled with tactical gross weight overages, led to many accidents. To offset the drastic imbalance of lift produced by a single rotor, blade system were designed to change pitch, flex, twist and change speed regardless of their position in relation to the wind. This "fully articulated" rotor head, however, resulted in a very unstable aircraft. These heads subjected helicopters to vibration, with high frequency vibrations producing serious stress. One solution was the design of two contra-rotating rotors. The tremendous torque created by a single rotor system was countered by the installation of a tail rotor. If the tail rotor failed; however, the aircraft would enter a spin, often with disastrous results.

One unique characteristic of helicopters is "ground effect". Under certain conditions, when the aircraft is near the ground in a hover or moving slowly, rotor downwash forms a thick air mass. A helicopter can slide off this cushion or if the ground is not smooth, the wash can break up, often providing some anxious moments for the crew.

Another phenomenon is "transitional lift" acquired as the aircraft increases forward speed, resulting in a sudden decrease in the power required to fly. The reverse occurs on landing when the aircraft slows and power must be added to maintain flight. If neither condition is present, the ability to lift heavy loads is greatly reduced, often resulting in mishaps.

Despite handling improvements brought about by sophisticated systems, helicopters remain inherently unstable. Unlike their fixed-wing contemporaries, helicopters demand the pilots full and constant attention. Improvements, including structural strengthening, dual hydraulics, weather radar and twin engines have helped to broaden the safety margin. Standard equipment comprising armor and self-sealing fuel tanks have dramatically minimized combat losses.

More helicopters were lost in Southeast Asia to operational, or non-hostile action, than in combat. Besides a cunning enemy, other factors not in the helicopter's favor were treacherous terrain, severe weather, maintenance difficulties, aircrew shortages and inadequate logistical support. Despite these obstacles, maintenance crews prevailed against all odds to maintain impressive aircraft availability rates. Another positive aspect was aircraft recovery using heavy helicopters fitted with a sling apparatus to return to bases downed aircraft that would have otherwise been lost. During the war 11,933 aircraft valued at 3.16 billion dollars were recovered.

Over 4,000 helicopters fell in Southeast Asia, more than half of them were UH-1 Hueys. Of these, the majority belonged to the U.S. Army. U.S. Air Force losses in SEA included fifty-one helicopters downed by ground fire, four to ground attack, seventeen to operational losses, two to unknown causes and a Sikorsky HH-53 shot down by a MiG. USAF helicopter losses during the Korean war included five downed by ground fire, five to "other combat", and seventeen to operational losses. Of a total of 1,724 Navy "major" helicopter mishaps from 1954 to 1976, 999 occurred during the Vietnam War years. One third of this figure were lost between 1967 to 1969. A total of thirteen USN and 270 USMC helicopter losses were attributed to hostile action during the Vietnam war. Also considered lost were approximately thirty-four CH-47 Chinook and 430 UH-1 Hueys left in Vietnam at the time of the communist takeover in April of 1975.

Early U.S. Air Force experiments with helicopters began with a number of Sikorsky H-5s during the late 1940s. This machine crashed at Wright-Patterson AFB in 1947. (via Larry Davis)

In 1983, American forces engaged in a rescue operation code named URGENT FURY. The invasion of Grenada embodied a force of 107 helicopters: eighty-five Army and twenty-two Marine. Of these, nine were destroyed, including six UH-60 Blackhawks, two AH-IT Cobras and one CH-46E. Most impressive was the Blackhawk's ability to withstand anti-aircraft fire. Of the thirty-two UH-60s committed to the invasion, ten were combat damaged with only one loss, all remaining losses were to accidents. The Blackhawks met or exceeded survivability and crashworthiness design specifications. One aircraft had forty-five bullet holes, punctured fuel tanks, holes in the main and tail rotors, much of the control instruments destroyed and a wounded pilot, yet the crew completed their mission!

Though inconsistencies are encountered (and expected) when researching official records, these statistics represent the commitment of the helicopter in the pursuit of American military goals. Often lost among these staggering figures are the harrowing accounts of aircrews who braved enemy fire and adverse conditions to carry out their mission. This book serves as testimony to the helicopter's affinity for danger while carrying out that mission.

This Sikorsky H03S-1 of Helicopter Utility Squadron One (HU-1) crashed aboard USS LOS ANGLES (CA-135) on 4 July 1951. Detachments from HU-1 were assigned to various ships for the air-sea rescue role. HU squadrons were redesignated Helicopter Combat Support Squadrons (HC) on 1 July 1965. (USN-Tailhook)

This Bell HTL-4 of Helicopter Utility Squadron Five (HU-5) carried an overall Gloss Sea Blue scheme. It crashed at Naval Auxiliary Air Station Brown Field outside San Diego, California on 27 March 1953. The tail rotor has been ripped from the aircraft and is on the ground next to it. (USN-Tailhook)

This Piasecki HUP-1 of HU-1 overturned on the deck edge elevator of USS RANGER (CVA-61) on 1 October 1959. The Navy accepted a total of 338 HUP-1s, with the first delivery of the type made in 1949 and the last delivered in 1954. (USN-Tailhook)

Troops of the 35th Quarter Master Company at Fort Carson, Colorado remove the tail section of a Bell H-13G Sioux from a crash site in the mountains on the Carson Reservation in May of 1956. Mules were the only means of removing the wreckage. (US Army)

(Left) The two fuel tanks from the H-13 are tied to a mule of the 35th QM Company. Behind the mule was a makeshift block and tackle rig hauled up to the mountain crash site by mule sled. (US Army)

This Sikorsky H03S-1 of HU-1 crashed at Brown Field, California on 26 March 1953. A total of ninety-one H03S-1s were delivered to the Navy. HU-1 was the last unit to use the type having phased them out in late 1954. The first loss of a helicopter during a rescue attempt was a H03S-1 of HU-1, Detachment 7 on 23 August 1950 over North Korea. (USN-Tailhook)

(Right) Marines inspect a Sikorsky H03S-1 of VMO-6 which crashed on a Korean hillside. H03S-ls of VMO-6 performed yeoman service as air ambulances and rescuing downed pilots. The first American aircraft to land at Kimpo Airfield after it was taken on 18 September 1950 was a VMO-6 H03S-1. The bracket assembly on the starboard fuselage mounted a 50 gallon saddle fuel tank to extend range. (National Archives)

This Army H-34A lifts an Italian Air Force Westland Huey during 1961. The H-34 was assigned to the 110th Aviation Company, Southern European Task Force. (US Army)

While the magesium skinning burned easily, not all H-34 crashes ended in fire. This aircraft was later repaired and placed back in service. (US Army)

Design changes in early production aircraft were often prompted by damage incurred during test phases. In its early development period, the Kaman H-43B development program was plagued with rotor-to-tail strikes during aft cyclic travel. During a demonstration at Stead AFB, NV in 1961,this type of accident occurred again and the severed tip of the inboard left tail fin is visible falling toward the firefighter. The wind was some thirty-eight knots, which may have contributed to the accident. (Kaman)

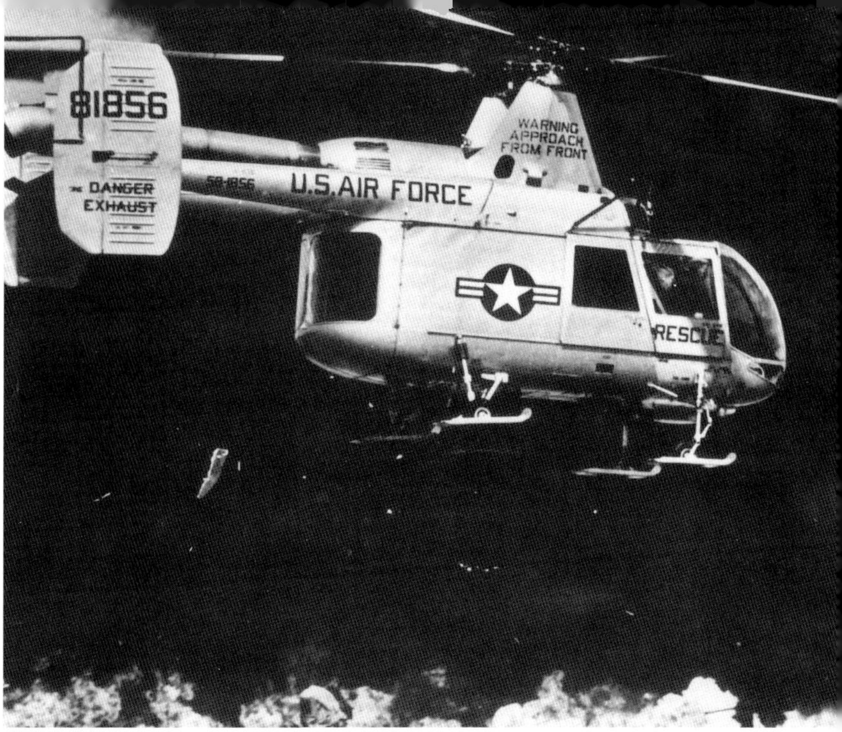

A Bell UH-1B gunship recovers another Bravo shot down during either late 1962 or early 1963. Due to aircraft shortages, early gunships retained their cargo hooks in the event they were needed to perform recovery operations. (Bell via Chenoweth)

Prior to sling recovery, an ARVN M-114 armored vehicle prepares to pull this gunship upright. Other accounts list this aircraft as a UH-1A shot down during the pivotal battle of Ap Bac during 1963. This battle under-scored the helicopter's vulnerability, despite armament. Official records indicate that no UH-1As were shot down during the battle. (Bell via Chenoweth)

During 1963 a small number of Sikorsky CH-37 Mojave helicopters were sent to Vietnam to fill the heavy helicopter recovery role. Downed aircraft could be recovered intact which reduced disassembly time in hostile areas plus subsequent reassembly time. By October of 1964, nine CH-37 were in-country. This CH-37B of the 611th Transportation Company at Vinh Long lifts a CH-21 Shawnee of the 57th Transportation Company which had been shot down near Ben Cat in 1964. (US Army)

This UH-1 Huey of the 187th Aviation Company took a direct hit during a Viet Cong attack at Tay Nihn during 1969. Fire fighting foam has completely covered the Huey, but not before the fire has completely gutted the aircraft. (George Sullivan)

After two days of work (under fire) maintenance personnel of Marine Aircraft Group 16 (MAG-16) successfully removed enough weight off this UH-34D Sea Horse so that it could be recovered by a Marine HR2S-1 on 31 October 1965. The crew removed the engine, engine compartment doors, tail section, rotors and transmission assembly. (USMC)

(Right) A downed UH-1H Huey of the 269th Aviation Battalion, 25th Infantry Division lies on its side after crash landing just outside the perimeter of an isolated mountain top camp near Cu Chi, South Vietnam during early April of 1967. (US Army)

This OH-13 Souix was put up on blocks after it was damaged in a hard landing. The aircraft was assigned to 1/9 Squadron, 1st Air Cavalry during 1969. (George Sullivan)

The remains of a UH-IB Huey rest alongside the runway at a base in Vietnam during 1962. It's believed that this Huey served with the Utility Tactical Training Company (UTTCO) which pioneered the helicopter gunship in Vietnam. The Bravo had an armament of 2.75 inch rocket tubes and mounts for four M60 machine guns. (Bell via Chenoweth)

Troops secure the crash site of a UH-IB of the 117th Aviation Company near Quinhon on 21 July 1964. The Huey (62-4574) was off loading troops during a combat assault in a hover when the tail rotor failed causing the aircraft to oscillate and roll over. The crew escaped unhurt. (Carl Vogel)

The 4,000 pound capacity cargo hook is visible in the "hell hole" of this UH-1B (62-4574). The hook was often used to recover other helicopters with sling rigging. (Carl Vogel)

This Bell AH-1G Hueycobra of A Battery, 2/20 Squadron, 1st Air Cavalry was shot down by enemy anti-aircraft fire during 1969. (George Sullivan)

A maintenance crew rigs a hoist to a downed CH-46A Sea Knight of HMM-165. The pilot landed the helicopter in the swift mountain river northwest of Chu Lai to extinguish the fire started by enemy fire. The aircraft was later retrieved from its watery landing pad. (USMC)

This UH-1D of B Company, 229th Aviation Battalion, 1st Air Cavalry was shredded by shrapnel from a rocket attack on Dan Tieng during 1969. (George Sullivan)

The meager remains of this Marine Sikorsky UH-34D Sea Horse of MAG-16 stands as mute testimony to the methodical enemy attack at the Marble Mountain Air Facility on 22 October 1965 which left nineteen helicopters destroyed and eleven heavily damaged. (USMC)

As Marines look on, a Boeing Vertol CH-46A Sea Knight, lifts a downed UH-34 Sea Horse from a landing zone during Operation COLORADO conducted in the Heip Duc area in August of 1966. (USMC)

This UH-IB Huey of the 1st Cavarly Division at An Khe was peppered with holes during a Viet Cong mortar attack on the "Golf Course" Airfield, Camp Radcliffe on 3 September 1966. (US Army)

Bearing mute testimony to the hazards of high-tension wires, the remains of this helicopter can barely be recognized as a UH-IB Huey. The incident occurred north of the Yakima Firing Range, Ellensburg, Washington in April of 1963. Hueys are now fitted with wire cutters to counter this hazard. (US Army)

Following a mid-air collision, rotor blades become deadly missiles as Marines escape this CH-46A of HMM-269 at LZ Crow, forty miles north of Hue during Operation HASTINGS on 15 July 1966. Armor plates added to protect vital engine components can be seen just forward of the exhaust on the rear rotor pylon. Four Sea Knights were lost in this valley prompting Marines to name the area 'helicopter valley". (USMC)

Rescuers rush to aid the crew members of these two UH-ID Hueys of the 13th Aviation Battalion which were badly damaged when one helicopter taking off crashed on top of another parked on the pad. The force of the impact is evident by the transmission and rotor mast (lying near the truck) which was torn from the descending chopper. (US Army)

(Above/Right) Helicopter Combat Support Squadron (HC)-7 flew Kaman UH-2A/B Seasprites under the callsign Iron Barnacle from offshore destroyers for combat SAR beginning in 1967. Tasked with the rescue of downed pilots in North Vietnam, the loss rate for Navy helicopters attempting over land rescues was very high. For every three aircrew rescues, two helicopters were lost. These losses were partly attributed to the lack of armor, self-sealing fuel tanks and heavy weapons on the SAR helos. This UH-2A (BuNo 149761) at DaNang during March of 1967, may have come under heavy fire from enemy guns above the Seasprite's position since the greatest concentration of hits (outlined with White spray paint) are in the rotors and upper engine areas. The cabin window has been taped over. (Tom Hansen)

A number of Bell OH-13 Sioux helicopters were used by the Ist Cavalry Division as scout aircraft before the Hughes OH-6A LOH was introduced in Vietnam. This OH-13S was damaged by mortar rounds during a VC attack on the "Golf Course" Airfield at Camp Radcliffe on 3 September 1966. By 1968, the Army had no more than twenty of the "Ultimate S" Model Sioux in Vietnam. (US Army)

The transmission and main rotor hub assembly lie alongside this UH-1 Huey at Quinhon following its recovery from the field. Aircraft of the 117th Aviation Company carried an unusual camouflage pattern during their first few years of operation in Vietnam. (Carl Vogel)

(Left) Besides the turbine engine,the Huey was also stripped of other vital components that could be used as spares for other unit helicopters. The copilot's collective lever, which controlled rotor blade pitch and incorporated the throttle, is visible next to the floor console. (Carl Vogel)

Most Army helicopters that received major structural damage, either in a crash or from battle damage in Vietnam were usually sent back to the United States via ship for depot level repairs. (Carl Vogel)

These two UH-IBs of the 119th Aviation Company at Pleiku, Vietnam collided in mid-air during 1964, resulting in the loss of both helos. One of the Hueys involved in the mid-air retained its rotor blades which often become lethal cleavers compounding damage, both to the aircraft and crews. (Carl Vogel)

(left) Post-crash studies and photographic evidence has shown that a large number of UH-1 Huey crashes resulted in the transmission being torn loose as a result of rotor blade strikes and failure of the tail boom after it was also struck by the runaway rotor blades and was sheared off. (Carl Vogel)

Crewmen strip a downed UH-IB of the 117th Aviation Company which crashed in Vietnam about 1964. Barely visible on the cabin roof of this Huey is a large, full color national insignia, part of the high visibility markings rarely seen after 1966. (Carl Vogel)

(Above & Left) A UH-IB Huey (62-4570) rests on the ground at Camp Goldbecg, Quinhon, Vietnam following a crash in 1964 that sheared off the tail boom. During its early stages of operations in Vietnam, the117th Aviation Company painted their aircraft in a unique camouflage of Tan and Black over Olive Drab. This crashed Huey appears well maintained and was probably repaired and placed back in service. (Carl Vogel)

A number of unmarked H-34s, camouflaged in Black and Olive Drab were used for covert missions in Cambodia under the code names Kingbee and Queenbee. This Queenbee Sikorsky crashed at Forward Operating Base (FOB) II at Kontum during 1966. Kontum was a major jumping off point for helicopters conducting high-risk clandestine missions "across the fence". (Tom Garcia)

A CH-37 hovers over a downed CH-21 while the crew hooks up the external sling to recover the helicopter. The area was secured by troops guarding the parimeter while the recovery operations are underway. Once the aircraft has been lifted out, another helicopter will take out the crew and security detail. (US Army)

A total of six USAF Green Hornet UH-Is were lost to operational causes. This UH-IF was destroyed on a training mission south of Kontum at White Castle Special Forces Camp during 1967. All 20th SOS Hueys were finished in Tan and Olive Drab over Light Gray. For its covert missions the aircraft were marked only with a partial serial number and the unit hornet insignia on the tail boom. (Tom Garcia)

Three disabled Marine HR2S-Is of HMM-462 await their fate at a Vietnam air base during March of 1967. The "Deuce" served primarily as a recovery platform capable of recovering both downed aircraft and helicopters until May of 1967. (Terry Love)

By the end of 1968, Boeing Vertol CH-47 Chinooks in Vietnam had retrieved over one billion dollars worth of aircraft that otherwise would have been abandoned or destroyed. This figure was more than twice the amount spent on the Chinook development program, making it one aircraft that paid for itself. This CH-47A lifts a UH-1B of the 175th AHC Outlaws. (Boeing Vertol)

A Marine UH-1E practicing combat assault landings collided with a CH-53 Sea Stallion at 500 feet over MACS New River, NC in June of 1967. The Huey severed the Sea Stallion's tail rotor pylon causing the aircraft to lose control. Thirteen Marines survived the crash which claimed twenty-two lives. Such statistics underscore the hazards of routine helicopter training missions. The Huey is at the bottom, the Sea Stallion at top right and severed tail rotor pylon is at top center). (USMC via Norman LaFountaine)

A Marine surveys the wreckage of the UH-1E of VMO-1, New River MACS, NC, after it had a mid-air with a CH-53A. (USMC via Norman LaFountaine)

Little but the nose section of the CH-53A remained as firefighters lay a layer of fire fighting foam to smother the still-burning wreckage. (USMC via Norman LaFountaine)

Resupply missions by helicopter into Khe Sanh during the seige were high risk as evidenced by this wrecked CH-53A Sea Stallion that was shot down in mid-February of 1968. A Marine removes the crew chief's M60 and ammo from the aircraft. The cockpit section of the Sea Stallion suffered a direct hit which proved disastrous for the pilots. (USMC)

A OH-6A of D Troop, 3rd Squadron, 4th Cavalry, minus its rotor blades, awaits a recovery helo after crashing in Vietnam. By the end of 1968 there were 542 OH-6s in-country despite losses of some 208 aircraft during that year. Army helos in Vietnam carried buzz numbers under the tailplanes. (via Bob Chenoweth)

Dustoff choppers in Vietnam often paid the price for their daring rescues under fire. This UH-ID of the 658th Medical Evacuation Company at Long Binh was brought down by rocket and small arms fire that hit the engine area in early August of 1967. The Red crosses served only as useful aiming points for enemy gunners. (US Army)

Fire has destroyed everything but the aft rotor pylon of this Marine CH-46A Sea Knight after it was hit during an NVA rocket attack at Dong Ha Air Base on 28 August 1967. (USMC)

(Left & Below Left) USMC helicopters in the northern regions of Vietnam suffered heavy losses to ground attack. Enemy mortar crews scored a direct hit on this Marine UH-1E gunship at Khe Sanh on 25 January 1968. This was the late Echo model Huey which had its counterpart in the Army's UH-1C. It was armed with nineteen shot 2.75 inch XM-200 rocket pods and a 7.62MM TAT-101 machine gun chin turret. (USMC)

Huey tail booms are fastened to the fuselage by four high-tension bolts. During a crash, the boom usually fails at a point just to the rear of that bulkhead. The hollow shaft visible above the tail boom is the tail rotor drive shaft. (Dave Fesmire)

Minus its tail fin and engine, the remains of this UH-1C (66-15083) of the 57th Assault Heilcopter Company lies derelict on the Kontum perimeter in 1968 following its third and final crash. (Irwin Baeder)

A CH-53 Sea Stallion of HMH-463 prepares to lift a CH-46 Sea Knight that had been damaged by enemy fire near Quang Tri in September of 1968. (USMC)

Two UH-1 Hueys of the lst Cavalry Division rest in sandbag revetments with their rotor retraining straps still attached after being sling loading back to Camp Evans following rocket hits in May of 1968. (US Army)

Despite high revetments, enemy gunners scored a direct rocket hit on this CH-47 Chinook of the 271st ASHC at Can Tho on 13 January 1969. The revetment did protect the OV-1 Mohawk in the adjacent parking area. Three 271st Chinooks were destroyed in the attack. (US Army)

LCOL Wilbur A. Middleton, Commander of the 212th Combat Support Battalion, surveys the tail section of a UH-1H that was torn off by an 82mm mortar round explosion during a VC attack at DaNang at the beginning of February 1968. Barrels filled with sand proved quite effective as revetment walls. (US Army)

(Left) A CH-46 Sea Knight lands to pick up the crew and passengers of another CH-46 which crashed when attempting to pick up wounded Marines from a hot landing zone. (USMC)

A motor whaleboat from the ASW support carrier USS HORNET (CVS-12) attempts to recover an SH-3A Sea King down in the open ocean as another Sea King hovers nearby. Flotation bags in the sponsons have been deployed for greater stability. (USN)

This Air Force Sikorsky H-5 crashed at Wright-Patterson Air Force Base, Dayton, Ohio during 1947.

During its lifetime, the Kaman H-43 Huskie underwent numerous tail section modifications. Following a rotor/tail strike accident in 1961, the H-43B underwent still further modifications.

This UH-34D Seahorse of HMM-162 was lost at sea following an engine failure off Vieques, Puerto Rico on 24 March 1967.

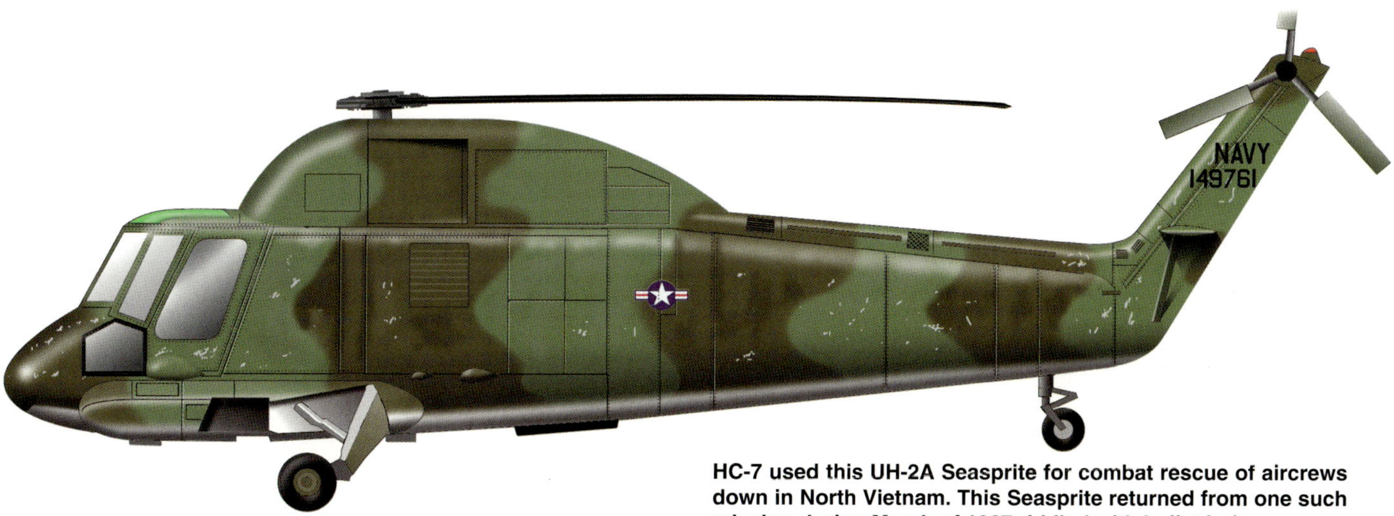

HC-7 used this UH-2A Seasprite for combat rescue of aircrews down in North Vietnam. This Seasprite returned from one such mission during March of 1967 riddled with bullet holes.

Crewmen aboard the repair, berthing, and messing barge YRBM-20 moored in the Mekong Delta fight a fire aboard a UH-1B of HA(L)-3, better known as the Sea Wolves. (US Army)

(Left) Aircraft downed in enemy territory were hastily retrieved by recovery crews who often faced hostile fire during operations, This CH-47A Chinook recovery team apparently wasted no time in snatching this downed UH-1H of the 176th AHC during 1969. (Dave Grieger)

This UH-1H (67-17397) of the 176th Assault Helicopter Company crashed in South Vietnam . The rotors are split down the center and only half of the rotor chord remains (Via Bob Chenoweth)

Unaware of his close call during a combat assault, this gunner of the 240th Assault Helicopter Company smiles for the camera. During the battle his helmet was ripped open and the visor cover torn away by an AK-47 round. The action took place near Bear Cat during 1971. (Author)

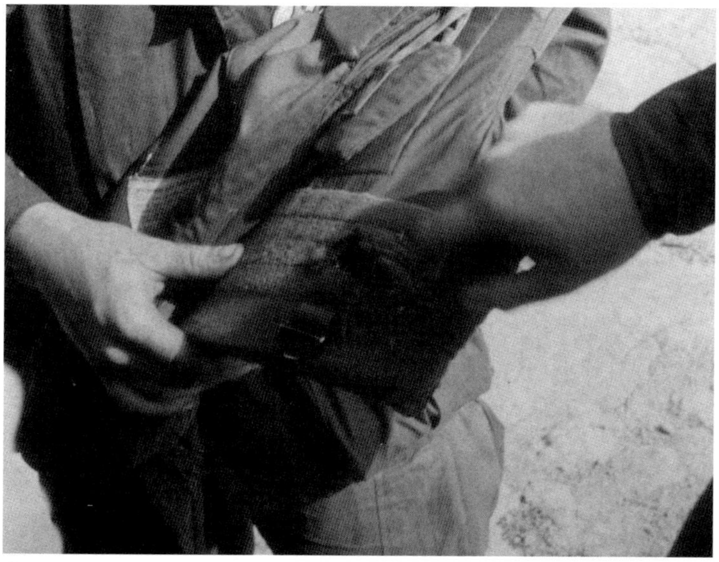

Enemy ground fire penetrated this Army helicoper crewman's chest plate armor. Better known as a "chicken plate', this ceramic body armor provided a large measure of protection to aircrews while engaged in their hazardous occupation. (via Author)

Fred Thompson, a UH-1 Huey pilot with the 174th Assault Helicopter Company received only minor injuries when an enemy round pierced and exited his flight helmet. The round went in though the front visor and exited just to the rear of the visor cover. (Fred Thompson)

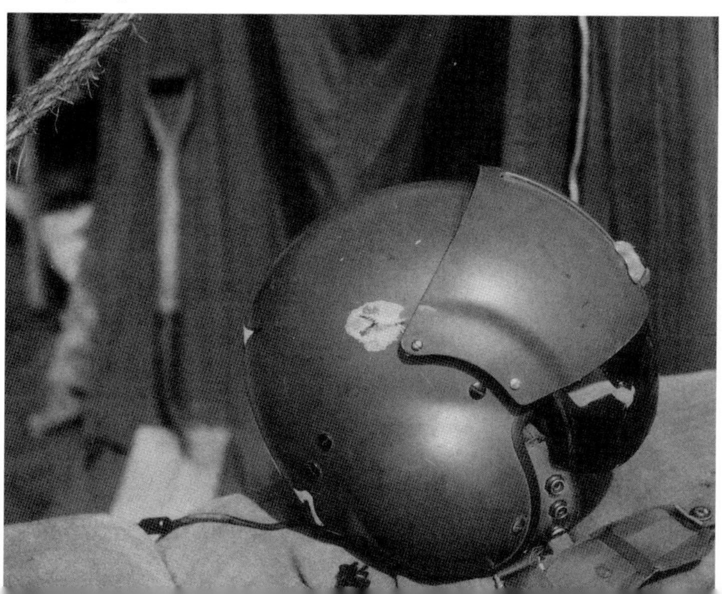

Members of the 101st Airborne Division at Camp Eagle examine the wreck of an AH-IG Hueycobra destroyed by an enemy rocket at El Toro Pad on 4 May 1970. (USN)

This UH-1B Huey of the 117th Avaition Company was destroyed in a crash at Quin Nhon, Vietnam during 1964.

This UH-1C carries pilot training markings and was intentionally destroyed while being used as a target for heat seeking missiles during missile testing.

A UH-1H of the Outlaws (175th Aviation Company) served as the Commanding Officer's aircraft until it was badly damaged by a VC mortar round during June of 1967.

This UH-1D of the 121st Assault Helicopter Company , known as the Soc Trang Tigers, had the call sign Tiger 27 and carried a Tiger's head on the copilot's door. It crashed during December of 1966 and was taken to the ramp at Tan Son Nhut to await its fate. (Terry Love)

This pair of wrecked UH-ID Hueys of the 254th Medical Company at Tan Son Nhut during November of 1966 attest to the high risk nature of the medical evacuation or Dustoff missions in Vietnam. (Terry Love)

Dark Horse Six was a UH-1H assigned to C Troop, 16th Cavalry, 1st Aviation Brigade and was armed with a .50 caliber machine gun in the door.

The 20th SOS used this UH-1P for cross border flights into Laos, where it was lost in combat.

A rotor blade from another helciopter sliced into the engine exhaust area and deep into the tail boom of this Huey which was also damaged by ground fire during early 1967. The tail rotor drive shaft and its hinged cover are visible on top of the tail boom. (Terry Love)

This UH-1D Huey of the 162nd Assault Helicopter Company, the Vultures, crashed in early 1967. Reportedly the aircraft impacted in the inverted position destroying the cockpit area. (Terry Love)

This UH-1C Huey gunship of D Troop, 3rd Squadron, 4th Cavalry, 25th Infantry Division, Centaurs, was being cannibalized following a rollover crash during late 1967. Usable parts from this aircraft will be used to keep other Hueys operational. (Terry Love)

The cause of the total destruction of this UH-ID of the135th AHC Cowboys is uncertain, although it appears to have occurred on the ground during 1961. Writing on the windshield states "do not touch - under investigation". Bullet 6 had been chalked on the nose denoting that the aircraft had been using the callsign Bullet at the time of the incident. (Terry Love)

A crew member stands on the transmission cover of this OH-6Aand hooks a sling to the cargo hook of a recovery UH-1H Huey of the Americal Division so that it can lift the OH-6A for removal to Chu Lai on 26 May 1970. The OH-6A went down at Landing Zone Siberia. (US Army)

A CH-54A Skycrane of the 478th Aviation Company (Heavy Helicopter) recovers a badly damaged CH-47A Chinook of the 228th ASHB. The aircraft was returned to the Ist Cavalry Division's Golf Course base at An Khe. (US Army)

All-around damage was inflicted to this UH-1H of the 57th AHC during a snapper attack at An Khe on 11 April 1970 Seventeen of the unit's twenty-three aircraft were either damaged or destroyed by the explosive charges. (Steve Hayduk)

The OH-6A LOH was introduced into the Vietnam conflict durin Fall of 1967. Earlier that year seventy-one OH-13s and forty-three OH-23s were moved from Europe and Korea to fill the vital scout helicopter role until the required 377 OH-6A could be deployed. The XM-21 minigun has been removed from this LOH downed in late June of 1967. (U.S. Army)

A ship board damage control fire party hoses down a UH-34D of HMM-163 to prevent spilled fuel from catching fire. It crashed aboard the amphibious assault ship USS OKINAWA (LPH-3) during Operation BALLISTIC CHARGE held in mid-October of 1967. (USMC)

Pilot error was responsible for the crash of this UH-1C of the Headquarters Company, 7th Squadron,17th Cavalry at Camp Evans, Pleiku, Vietnam during December of 1967. The Snoopy character (wielding crossed sabers) on the pilot's door was a popular mascot that appeared on a great many aircraft of the era. (Bob Steinbrunn)

(Left) Synchronized-shaft failure was one problem which plagued early variants of the Sea Knight like this CH-46D. In early December of 1967 the sync-shafts failed on this Sea Knight of HMM-162 during a training mission at MCAF New River, NC. The crash caused the rear rotor to cut the helo in half killing the crew of three. (Norman LaFountaine)

This crash damaged UH-ID was parked in a dump surrounded by other damaged UH-1 Hueys in Vietnam during 1967. Although written off, these aircraft often served as a source of spare parts for other UH-1s. (Bell via Bob Chenoweth)

This UH-1H Huey (67-17379) of the 176th Assault Helicopter Company, the Minutemen, flew over a demolition area just as it blew up during 1969. The aircraft was at an altitude of fifty feet at the time of the explosion. Reportedly the crew survived with only back injures. (Dave Grieger)

Crewmen from the 180th ASHC survey the damage to their CH-47A after it was shot down while conducting a sling load delivery during 1968. Battle damage is visible just to the rear of the pilot's door. (Bob Steinbrunn)

This UH-1H Huey of the 92nd Assault Helicopter Company was one of a number of helicopters heavily damaged by a Viet Cong sapper attack at Dong Ba Thin during 1970. (Ray Fletcher)

The rotor head, cabin doors and miniguns have been removed from this UH-1P of the 20th SOS Green Hornets that was shot down during August of 1970. The remaining seven-shot LAU-59/A (XM-157) 2.75 inch rocket pods were salvaged later. The crew survived the crash uninjured and was rescued. (Gene Cole)

Hughes OH-6A Light Observation Helicopters (LOHs) were used to fly low and slow reconnaissance missions designed to draw enemy fire forcing the enemy to reveal their positions. This LOH of 1/9 Cavalry, 1st Cavalry Division was shot down near Camp Bearcat during 1971 while conducting just this sort of mission. (Author)

This CH-37 Mojave crashed while attempting to recover a crashed UH-1B Huey from a clearing north of Bien Hoa Air Base on 23 September 1965/ (US Army via Jim Mesko)

Crewmen scramble to evacuate the crew of this 176th AHC UH-1H shot down near Binh Sonh on 30 December 1968. The Huey lost its tail rotor after receiving ground fire while on a "sniffer" mission.(Dave Grieger)

This CH-47 of the 178th ASHC Boxcars was in a fifty foot hover when it took six ground fire hits in the aft transmission area during the Summer of 1969. While on fire it was flown to Tam Ky where it crash-landed and was destroyed. (Dave Grieger)

A CH-54 Skycrane heavy lift helicopter recovers a CH-47A Chinook. The rope coming off the rear of the CH-47 went to a drag chute that helped keep the Chinook steady as it was being transported. The Skycrane was the only helicopter in Vietnam capable of recovering a Chinook. (Sikorsky)

This UH-1H Huey of the 92nd Assault Helicopter Company was shot down on Highway I, the main north/south road in South Vietnam, just outside Dong Ba Thin during 1970. The force of the impact tore away the tail boom and the landing skids and crushed the cockpit. (Ray Fletcher)

Vietnam's high-density altitude coupled with heavy loads often spelled disaster for gunships such as this UH-1M of the 25th Infantry Division which never cleared the pad at Tay Ninh during August of 1969. Full ordnance loads such as the rockets still in the XM-200 pods on this Huey posed a serious post-crash threat. (Mike Stratton)

Combat operations were not the only area where helicopters were lost. Training accidents were not uncommon among the ranks of fledging Army aviators leading to the loss of a number of helicopters. The basic trainer for helo pilots was the TH-55, this example ended its career on its side in a farmers field. (U.S. Army)

This Huey did fare well following a mid-air collisionv with another Huey during late 1971. Luckily, both aircraft had already inserted their troops just prior to the mishap. Both aircraft involved in the accident were UH-1Hs of the 162nd Assault Helicopter Company. (Dave Fesmire)

Considering the amount of helicopter traffic during a combat operation in Vietnam, mid-air collisions were not uncommon and often resulted in the crash of both aircraft. This UH-IH of the 162nd AHC collided with another Huey following a troop insertion during late 1971. (Dave Fesmire))

In 1966 the 174th AHC received permission to use the shark-mouth design from BGEN (Ret.) Robert L. Scott, who commanded the famed 23rd Fighter Group which had taken over from the Flying Tigers during WW II. Their P-40s carried a shark mouth similar to that carried on this UH-IC gunship during early1971. (Fred Thompson)

A pair of OH-6A LOH scout helicopters rest in the bone yard at Cu Chi base camp during the Summer of 1969. Both aircraft were shot down while trolling for enemy fire. (Dennis Dennison)

The Boeing Vertol CH-47 Chinook became the primary vehicle for recovering downed aircraft in Vietnam. This CH-47A of the 330th Transportation Company kept a running tally of aircraft recovered with Black silhouettes. The total as of November of 1967 was 167. (Terry Love)

A crew chief proudly points out the running tally of downed aircraft recovered by *"Rice Paddy Daddy"*, a UH-ID of the 56th Transportation Company. The tally includes both helicopters and fixed wing aircraft. (Bell)

An AH-1G HueyCobra of the 17th Cavalry rests in the bone yard at Cu Chi base camp during the summer of 1969, awaiting its turn to by stripped of all usable parts. Also in the yard are an AH-1G of the 20th ARA and a UH-1H of the 116th AHC. (Dennis Dennison)

This UH-1H (66-15569) of the 57th Assault Helicopter Company Gladiators was one of the aircraft destroyed during a Viet Cong sapper attack at An Khe. (Steve Hayduk)

(Left) A damaged Huey is recovered by a CH-47A Chinook using the underfuselage cargo hook. During this type of operation the crew chief would watch through the open hatch in the aircraft's belly to make sure the sling load was not swinging. (Via Jim Mesko)

An SH-3A Sea King of Helicopter Combat Support Squadron (HC)-6 based at the Naval Amphibious Base, Little Creek, Va, taxis toward shore some ten miles away. The Sea King experienced engine failure on 27 October 1969 forcing it down at sea. (USN)

The potential for damage incurred by support aircraft in close combat quarters was often averted by pilot skill as illustrated here. This UH-IH of the 240th AHC lifts out of the LZ with a load of troops as a"Mad Dog" gunship places suppressive rocket fire a mere 100 feet off the "slick's" right side. (Author)

This UH-1H of the 174th Assault Helicopter company made a successul crash landing after being shot down during August of 1970. The crew has bailed out leaving the rotors turning as fire breaks out in the engine compartment. The turning rotors only intensify the fire and it eventually destroyed the Huey. (Fred Thompson)

Super Scenic Cruiser was a UH-1H belonging to the 240th AHC Greyhounds which crashed near Camp Bear Cat during September of 1971 while flying a Thai combat support mission. (Author)

Ground fire pierced the nose compartment door of Darkhorse 6, a UH-IH flown by the commander of C Troop/16th Cavalry. The Huey was on station over the U Minh Forest on 9 December 1971 directing air strikes against enemy forces that shot down two OH-6As and damaged four Hueys. (Dave Fesmire)

Marine salvage crews bring in a crane to lift this CH-46 Seaknight off the roof of the 1st Marine Air Wing Post Office building during 1968. The crewmen on top of the helo are attaching cables to the lift points while others are starting to remove the rear rotors. (Lawrence Gibbs)

This LAMPS SH-2D Seasprite was destroyed in a crash during the mid-1970s.

This SH-3A of HC-1 suffered major damage when it flipped over on USS CORAL SEA in the Spring of 1972.

Training missions in the U.S. also took their toll of Army helicopters. After surviving the Vietnam war, this UH-IH met its fate on a mountain side near Fort Carson, Colorado during late 1975. (U.S. Army)

The CH-47C of the 213th Aviation Company landed safely at Daesa Ri, Korea during 1979 despite the severe damage that resulted from a seized gearbox which exploded. (Bod Dollar via Pete Harlem)

A CH-47A Chinook airframe impacts following a drop from a 250 foot tower at the NASA Langley Research Center's Impact Dynamics Research Facility. Airframes were commonly fitted with crew dummies and simulated fuel loads for accurate crashworthiness tests. This facility has been in use for such tests since the early 1970s and was originallly built to support America's lunar-landing. (NASA)

A Marine CH-46 Sea Knight circles protectively overhead while rigging crews prepare another CH-46A of HMM-265 for recovery following a combat crash during 1968. (USN-Tailhook)

This UH-1D of the 173rd AHC Robin Hoods at Lai Khe served as the mount for Medal of Honor recipient SP/4 Gary Wetzel who was severely wounded when his "slick" was repeatedly mauled by rocket and machine gun fire near Can Giuoc on 8 January 1968. All fourteen Hueys involved in the combat assault were hit by ground fire with three being shot down. (Gary Wetzel)

Specialist 5 Mike Magar points out damage from .51 caliber machine gun fire to the fuselage of this UH-IH of C Troop, 16th Cavalry, call sign Darkhorse, during 1972. The top surfaces of synchronized elevators of Army Hueys were painted Orange as a spotting aid to orbiting aircraft. Hueys assigned to units in the 1st Aviation Brigade carried their battalion geometric symbol just forward of the elevator. Besides the damage to the fuselage, the tail rotor gearbox was shattered by the .51 caiber anti-aircraft machine gun fire. (Dave Fesmire)

(Left) The American crew of this Vietnamese Air Force UH-1H abandons the Huey after it was shot down in August of 1972. Helicopters seldom ventured into enemy territory alone so that in the event of an emergency, an aircraft in company could pick up downed crews. (Mike Lecroy)

This USAF HH-53C Jolly Green rescue helicopter of the Aerospace Rescue and Recovery Service was one of twelve aircraft destroyed or damaged by North Vietnamese Army (NVA) rocket attacks at Da Nang Air Base in August of 1972. (USAF)

The cockpit area of this Vietnamese Air Force (VNAF) UH-IH was demolished following a crash in late 1971. The instrument panel is at far left and the pilot's collective lever is at lower center. Another H model wreck rests in the background. U.S. maintenance advisory teams at Bien Hoa waged a losing battle in the effort to train VNAF crews. Before long the VNAF graveyard at Bien Hoa was amply supplied with Hueys. (Author)

(Left) *THE IRON BUTTERFLY* was a UH-IH Dustoff Huey downed in August of 1969. The rigging used to secure the rotor blades during the sling recovery operation conducted by a 243rd ASHC CH-47A Chinook is still in place. (Koening)

This UH-ID Huey of the 335th Assault Helicopter Company Cowboys was destroyed in a crash during February of 1967. (Terry Love)

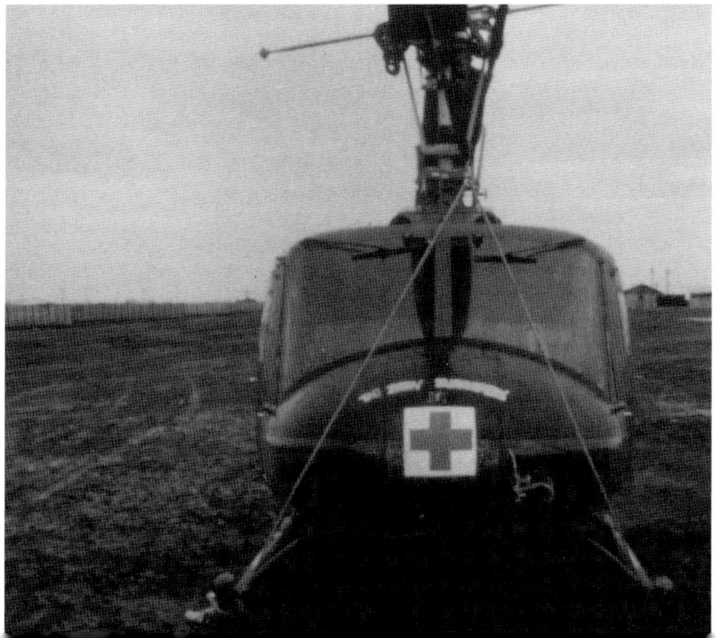

This HH-3E was dropped into Da Nang Harbor by an Army CH-37 Skycrane while being transported. The aircraft was assigned to the 37th ARRS and was the 1st production HH-3E. (E.E. Hatch via Al Adcock)

This UH-1 Huey crashed and sank in the water not far from the main gate of the 9th Infantry Division base at Can Tho, Vietnam during 1969. (Robert F. Fischer)

This South Vietnamese Air Force CH-34 was damaged by shrapnel during a Viet Cong mortar attack on Binh Thuy Air Base during the late 1960s. (USAF)

An HH-3E on the ramp at Tan Son Nhut Air Base just outside Saigon minus its rotors while undergoing repairs following a crash. The HH-3E saw widespread service in the air crew rescue role. (Terry Love)

(Above and Left) These UH-1s were all retrieved from the bush and taken to Camp Holloway (Pleiku). All three aircraft were shot down and were assigned to the 57th Assault Helicopter Company, 1st Aviation Brigade. (Phillip Friddell)

(Below and Right) The Marines used this area at Da Nang Air Base as a helicopter graveyard during March of 1967. Included in the boneyard were CH-37s, CH-46, CH-34s,UH-1s and even a few O-1E fixed wing aircraft. (Via Terry Love)

This CH-34A (54-3034) of the 361st Aviation Company at Wertheim, West Germany collapsed the main landing gear in a hard landing accident during September of 1968. (Terry Love)

(Right) This destroyed UH-1P Huey was armed with a Minigun in the forward cabin. The aircraft was assigned to the 20th Special Operations Squadron and flown on missions into Laos. (Jim Ball)

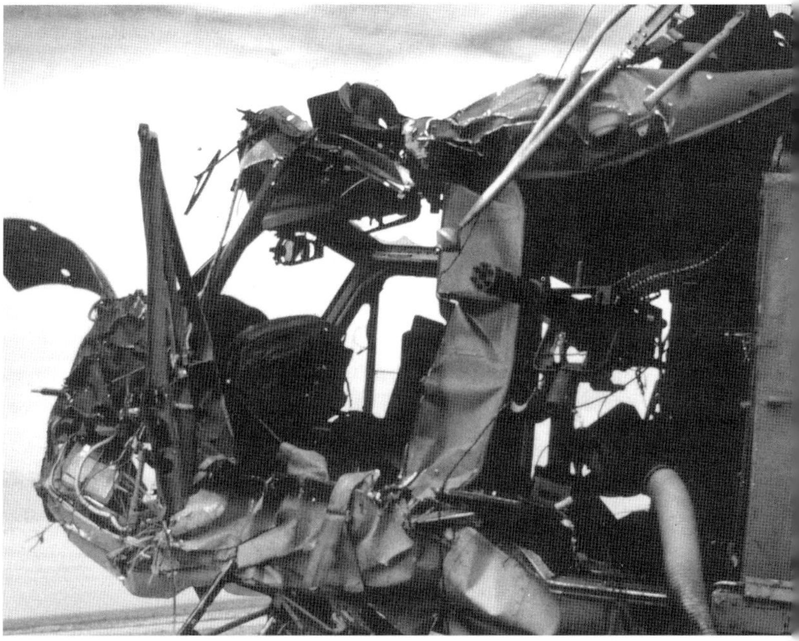

A number of war-weary Hueys proved their usefulness right to the end. This UH-IC (714091), still wearing pilot training markings, was used for missile tests. The aircraft was placed on a platform, with the engine running, to test the effectiveness of various heat-seeking anti-aircraft missles during the late 1970s. An inbound missle is visible emitting a smoke trail just over the tall boom. (Author)

An SH-3A Sea King came to rest over the catwalk of the attack carrier USS CORAL SEA (CVA-43) during operations in the Pacific Ocean. The helo jumped the deck edge combing and flipped over on 21 May 1972. A deck crane was later used to right the aircraft which was assigned to HC-1. The aircraft carried a buzzard painted on the engine intake screen above the cockpit. (USN)

This Army National Guard UH-IH Huey was totally destroyed in a training accident during the early 1970s . The aircraft nosed over into a field and rolled inverted, crushing the cockpit area . The fate of the crew is unknown. (US Army)

A US Navy motor whaleboat moves in to rescue the crew of this Vietnamese Air Force UH-1H ditched into the Tonkin Gulf during the evacuation of Saigon in April of 1975. (National Archives)

A VNAF UH-1H is pushed over the side of the USS BLUE RIDGE (LCC-19) to make room for other helicopters during the evacuation of Saigon on 29 April 1975. A total of fifteen VNAF helos landed aboard BLUE RIDGE while some 430 Hueys were left in Vietnam at the takeover. (USN)

A Vietnamese Air Force (VNAF) UH-1H lands inverted in the ocean after being pushed over the side of the USS BLUE RIDGE to make room for incoming helicopters during the evacuation of Saigon on 29 April 1975. (USN

The crew of this SH-3A of HS-6 awaits rescue after ditching their Sea King on 22 January 1976. Ditching procedure called for the hatches to be opened or removed and for the flotation bags stored on the landing gear sponsons to be deployed. (USN-Tailhook)

While a motor whaleboat stands by to assist, a deck edge crane hoists an SH-3H of HS-12 aboard USS TRIPOLI (LPH-10) after being forced to ditch in the open sea during 1982. The helicopter was kept afloat by its hull and the deployed flotation bags on the main landing gear sponsons. (USN-Tailhook)

A total of six Army UH-60 Sikorsky Blackhawk helicopters were damaged or destroyed during Operation URGENT FURY, the invasion of the island of Grenada. Three of the four pictured here were damaged in the air assault on Calivigny Barracks on 27 October 1983. (DOD)

One Marine CH-46E Sea Knight unit was deployed as part of Operation URGENT FURY. HMM-261, based at Marine Corps Air Station New River, NC lost one of its CH-46Es when it was disabled by enemy fire during a Ranger insertion and abandoned on Grand Anse Beach. It was then strafed by an AC-130H to prevent enemy salvage. (DOD)

A Boeing Vertol CH-47A Chinook airframe impacts following a drop from a 250 foot tower at NASA's Langley Research Center's Impact Dynamics Research Facility. Airframes such as this were commonly fitted with crew dummies and simulated fuel loads for accurate crashworthiness tests. "Belly Whopper" tests provided valuable information on Chinook crash impact for use in survivability improvements. (NASA)

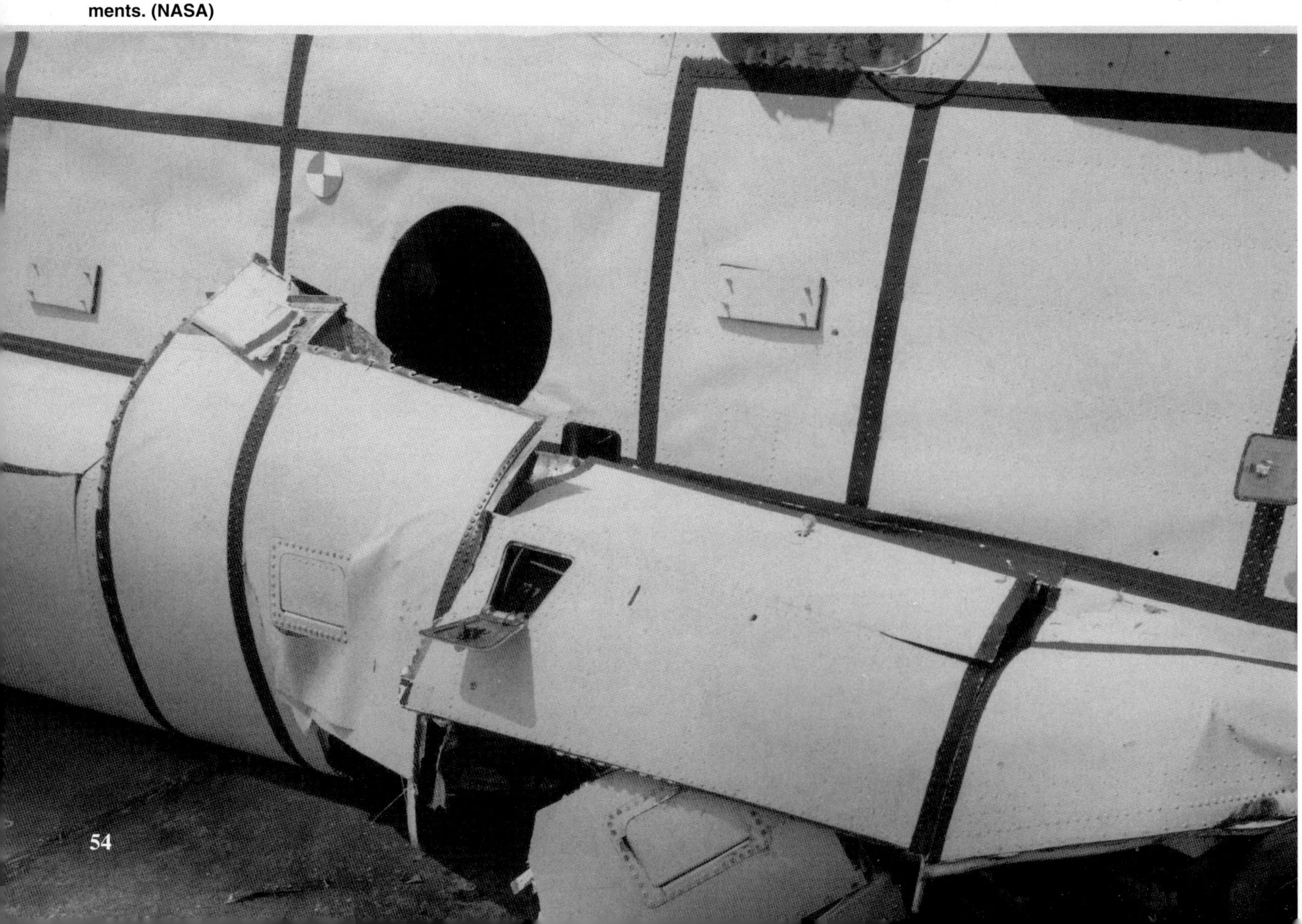

This CH-53 Sea Stallion, based aboard the USS SAIPAN (LHA-2), lost one of its main wheels in flight. The helo circled the ship to burn down fuel and finally was recovered aboard using a cart piled high with mattresses as a replacement for the lost wheel. (Nick Waters III)

Commissioned on 31 July 1973, HSL-33 (Helicopter Anti-submarine Squadron [Light]) was the Navy's first unit outfitted with the Light Airborne Multi-purpose Systems (LAMPS) in the Pacific Fleet. This LAMPS-equipped Kaman SH-2D (BuNo 149015) crashed during the 1970s. Visible underneath the aircraft is an auxiliary fuel tank. The magnetic anomoly detector (MAD) gear was carried on the starboard pylon. (USN-Tailhook)

A Sikorsky SH-60 Seahawk struck the water during a low-level flyby for a Navy frigate during exercises held in 1988. Pieces of the horizontal stabilizers and tail rotor are visible behind the tumbling helo. (USN)

U.S Forces examine a burned out Iraqi Air Force Mil Mi-8 Hip helicopter on an airfield in southern Iraq. Many Iraqi helos were caught out in the open and destroyed by cluster bombs. (USAF)

This burned out hulk was all that remained of an Iraqi Air Force Mi-8 Hip helicopter that was caught out in the open on Tallil Air Base by attacking U.S. strike aircraft. (USAF)